MW01234728

This Journal Belongs To
